Table of Contents

INTRODUCTION

You've heard of aeroponics before, but what about fogponics? Sometimes called the next phase in aeroponics, fogponics is a hydroponic system that suspends plant roots in air and supplies them with nutrients via water droplets that are so small, they qualify as vapor. Here's how it works.

If you're into hydroponics, regardless of the system you use, you're no stranger to water and nutrient management. From DWC (deep water culture), to NFT (nutrient film technique), to aquaponic systems, the "hydro" in hydroponics gets top billing for a reason. Soilless gardening relies on water as the primary vehicle used to deliver nutrients to hungry plant roots.

CHAPTER ONE
Fogponics

Fogponics is an offshoot of aeroponics. In this indoor gardening practice, plants are suspended in enclosed systems with no soil or growing medium. Instead, a fog or mist of water and nutrients is pumped into the closed system to give the roots constant exposure to the nutrients the plant needs to grow.

The theory behind fogponics is that plants are best able to absorb particulate nutrients in the 1 to 25 micrometer (μm) range. Therefore, plants in a fog or mist mixture can better absorb nutrients for more efficient gardening and less waste.

MaximumYield explains Fogponics

Fogponics is an advancement on aeroponic gardening techniques. With fogponics, though, an opaque crate, bucket, or tray is used to seal in the plants' roots. Holes are cut in the top to accommodate the net cups in which the plants will grow, and one more hole is cut in the side or bottom for an electric fogger.

The fogger introduces a mist of nutrient-water mixture into the closed system, and the plants' roots are constantly exposed to nutrients at particulate sizes that they easily absorb and take up.

Because there is no need for soil, fogponics offers a sustainable and efficient means to grow vegetables and other plants indoors. Fogponics is also more efficient than hydroponics in that all the water vapor and nutrients for the system are trapped within the system and nothing is lost to evaporation.

Fogponics does require regular use of electric power to introduce a mist or fog into the system. When setting up a fogponics system, indoor horticulturalists must consider a backup power supply during a power outage, as it could be devastating for their plants. Likewise, the system must be regularly maintained to effectively deliver nutrients.

Fogponics, however, allows growers to create very specific nutrient-water mixtures to create more suitable growing conditions for certain plants. Some vegetables, such as tomatoes, respond very well to fogponics. Furthermore, with fogponics it is easier to evenly distribute and disperse nutrients and water to all roots equally.

Figuring Out Fogponics

What may not be as obvious is the role oxygen plays. Leaves and stems help carry out photosynthesis, producing oxygen as a byproduct, but at the root level, plants need oxygen from the soil to help break down the food they need to survive.

Plants rely on the efficient delivery of water, nutrients and oxygen to grow to their full potential. And one system that delivers an ideal mix of all three of these essential items is fogponics.

Let's take a look at a few different hydroponic options to help us understand what makes fogponics so special.

Media-based Hydroponic Systems

In a basic hydroponic set-up, media is a substitute for soil. There are lots of different types of hydroponic media, but they all tend to be chemically inert, heavy enough to anchor plant roots and porous enough to allow for good oxygen access and water/nutrient distribution.

There are some drawbacks to using media, though. Media can clog feed lines, entangle roots, harbor bacteria, algae or fungi, or lose porosity over time.

In some systems, switching from one type of media to another can be a problem, too.

Immersion Systems

In an immersion system, delicate roots remain in a liquid environment most if not all of the time, and definitely throughout a plant's early development.

The anchoring media is replaced by a basket or other simple support structure that makes roots easy to see and easier to remove when or if plants are eventually transferred to soil or another growing environment.

When roots are suspended in water more than intermittently, they need special treatment in the form of supplemental oxygen.

The gaps and pores in conventional growing media make oxygen readily available to roots, and conventional hydroponic systems also include processes that remove and replace water and nutrients regularly, giving plants increased exposure to the air.

In submersion systems, oxygen is added through the use of an air stone, the same way oxygenated water is supplied to fish in an aquarium.

Because plants are in the water/nutrient mixture all of the time, pH fluctuations and temperature instabilities can happen suddenly and be devastating.

Going Airborne

Aeroponics approaches things differently by bringing water/nutrients to plant roots suspended in air instead of providing oxygen to plant roots suspended in water. The basic idea is pretty simple. Plants are installed on a platform using a support collar.

The roots dangle below into a box structure containing a network of spray nozzles that deliver nutrients to them on a timed schedule. Excess nutrients are recycled and reused.

During the early development of aeroponics, some interesting things came to light. One of the most significant is that the increased access to oxygen in an aeroponic system enhances plant growth a lot.

Another is that the continuous delivery of nutrients and oxygen in lavish amounts is an efficient way to farm.

Plants don't waste energy searching for food, and since there is no real competition for nutrients, and no incidental reduction in oxygen supply, plants can be grown in close proximity to one another, saving space, time and money. Sounds pretty good.

Fogponics: A Better Mousetrap

Fogponics is a form of aeroponics, sometimes called the next phase in aeroponic technology. Think of it as a refinement, but a significant one.

It uses the same basic concept as aeroponics—the practice of suspending plant roots in the air of a grow box and supplying water/nutrient droplets on a timed schedule.

In fogponics, though, the droplets are so tiny, they qualify as vapor. And in this case, smaller is better.

Plants absorb elements most efficiently in the range of about 1-25 micrometers (μm), and though foggers work somewhat differently depending on the manufacturer and model, a fogging system will supply vapor at between 5 and 30 μm.

While aeroponic systems supply nutrients to plant roots, foggers deliver vapor to stems and leaf nodes as well as roots, resulting in more efficient nutrient uptake and healthier plants overall.

Plants absorb droplets well, but because they absorb vapor better and faster, this can be good news for growers.

There's less waste, good space conservation and less opportunity for error, especially during root feeding.

How can you make a mistake spritzing plant roots to feed them? Aeroponic systems rely on precisely spaced nozzles to spray the interior of a grow box evenly.

Sometimes there are voids, or plant roots can become so thick they create obstructions.

Clogged nozzles can also fail, killing plants that are locked out of water and nutrients. These types of problems may not be detected until it's too late.

However, with fogging systems, wider and more comprehensive dispersion is much easier to accomplish, and growers are more likely to get to all the roots evenly and consistently.

This type of delivery system may also change the nature of the roots on some plants by encouraging smaller, hair-like growths that are even better at taking advantage of vapor-based feeding.

The Facts on Fogponics

Although fogponics has been around for a few years already, many gardeners remain unfamiliar with this method of growing.

Fogponics is the latest advancement in aeroponics and is an effective way to grow whatever plants you choose inside what could rightfully be deemed a fog chamber.

Inside this chamber is where the roots of the plants are suspended and exposed to the magic of the fog's hydration and nutrition.

To give you an idea of what it looks like, a dark plastic crate with a sturdy lid is required. If you don't have a dark crate, you can always do what the Stones did and paint it black. The point is, you don't want light seeping in.

This crate will serve as your growing chamber. Next, you'll need to cut out enough properly sized circles in the lid, which will hold the net cups your plants will be growing in. Six holes on top should do it. An additional hole is required to hold the small electric fogger.

Electric foggers, or mist makers as they are also known, are available in single-node, three-node and five-node varieties, to my knowledge.

Their job is to vibrate at 2 MHz (2 million times per second), transforming water and added nutrients into 100% humidity (fog), thus continually watering and feeding a plant's roots in a nutrient-rich cloud of mist.

There isn't time to tell you how to completely build a fogponics chamber; my intent here is to merely introduce you to the concept and give you a mental picture. There are several instructional videos available online that discuss the subject of construction and the maintenance of the system.

The results of growing this way are impressive, to say the least. When done correctly, you can expect greater yields while using less energy, and you will have a system that will make you the envy of traditional growers everywhere—even your hydro friends.

Tips for Growing Plants Using Fogponics

Once constructed, some growers use a small fan to circulate the mist, while others say even a small fan is too strong and only forces the fog out through any gaps that may be present. Another consideration is the fogger itself.

A single-node fogger is sufficient, but it all depends on the volume of plants you're growing and the size of your fog chamber.

General-purpose foggers—not designed specifically for horticultural purpose—typically come with LED lights around them, which you'll want to paint over as well.

Black paint that you would use to paint a model car or airplane will do. You can buy foggers specifically built for fogponics, but the cost can be five times higher than the general-purpose versions with the LEDs.

One pitfall is that these foggers can quit working when used around the clock. Leaving a fogger powered off for even half a day can cause severe damage to plant roots as they can dry out.

I highly recommend you incorporate an electronic timer that turns the fogger off for a while and then back on, doubling the lifespan of the mister and reducing the risk of a malfunction.

Commercial growers will want to use foggers designed specifically for fogponics, as they will likely be built for constant use (i.e. stronger).

Advantages of Growing Plants Using Fogponics

Outstanding yields

Less space required

Reduced energy costs

Controlled indoor environment

Easy to construct and maintain

Mesmerizing and really fun to play with

There you have it. Fogponics, a.k.a. Aeroponics 2.0, is just waiting for you to give it a try. Best of luck!

CHAPTER TWO
What is Aeroponics

So let's get to the true root meaning of what aeroponics is. The word Aeroponics is derived from two Latin words meaning aero (air) and ponic (work), air at work. In other words, the whole plant, roots and all, are suspended in mid air. Simple put the roots are not covered, not buried, not submerged in any form of liquid or matter, they are revealed openly in thin air.

It's so simple and basic. Why do so many complicate the issue? Because, they have their own agenda to make or they want to profit from it. Please, don't be taken by such false advertisement.

Good! We got that out of the way. We will return to the definition later. But for now let's focus on what aeroponics really is.

Getting right to the point, Aeroponics is the fastest way to grow plants than any other growing method. Aeroponics is the process of growing plants in thin air where its roots are exposed to a misty environment. No soil or aggregate medium is used to support the plant.

Aeroponic is an advance method of developing plant growth, flowering, fruiting and health for most plant species and cultivators. Aeroponics is a method of growing plants suspended (hanging) in mid air without the use of soil or any medium. The bare exposed roots get all of their nutrients from the environment it's in. There are two forms of aeroponics: High Pressure Aeroponics (HPA) and Low Pressure Aeroponics (LPA).

Carbon dioxide in the air is necessary for healthy plant growth. However, oxygen is needed for roots to absorb the nutrients that surround them. True aeroponics is conducted in air enriched with micro-droplets of nutrient water.

Did you know?

The most popular aeroponics system sold on the market is a lie. It's not aeroponics. The Aeroponics system can grow herbs, vegetables and flowers continuously to allow the grower to obtain yields like never before.

Aeroponics is perfect for indoor gardening, grow rooms and greenhouses. However, it has been used successfully for outdoor growing of large or tall plants.

Which Aeroponics are you talking about? High or Low

This opens another issue about aeroponics. There are two types or forms of aeroponic systems. The first one is HPA (High Pressure Aeroponics), and the other is LPA (Low Pressure Aeroponics).

HPAs are considered to be True Aeroponics and was used by NASA to grow vegetables. It also is the most expensive and the most complicated growing system to build. However, HPAs use less resources for plant grow: 98% less water, 60% less fertilizer, and 100% less pesticides (no pesticides), all supported by NASA laboratory studies.

LPAs are lower cost system. LPA systems are the most common used and built by DIYers.

High Pressure Aeroponics is True

The HPA (True Aeroponics) system was revolutionized by NASA in the 1990's by reporting it as the most efficient way to grow plants in space. Studies have shown many benefits of growing plants with aeroponic techniques on both Earth and in space.

HPA systems must operate at a high pressure, normally above 80 PSI, but ideally at 100 PSI. The high pressure is used to atomize the nutrient water through a small orifice (hole) to create water droplets of 50 microns or less in diameter, in other words a fine mist like hair spray.

One micron is one-millionth of a meter. The average diameter of human hair is 80 microns. So we are talking about a really tiny water drop.

HPA also must run on a much accurate time cycle. HPAs might run 1 to 5 seconds on, and then off for three to five minutes. Specific components are required in controlling the timing interval and creating the proper size mist.

The basic components of a HPA are as follows:

1. High-Pressure water pump

2. Pre-Pressurize Accumulator Tank

3. Electrical-Solenoid hooked to an adjustable relay timer

4. Pressure switch

5. Mister nozzles

Low Pressure Aeroponics is Cheaper

LPA systems use a standard mag drive pump couple to some PVC or tubing, and a few miniature sprinkler heads. The water spray from an LPA sprinkler head has large droplets that drown the plant roots.

LPAs generally run the pump 24 hours and 7 days a week, continually wetting the roots. This works well, and is cheap and easy to build. However they are not as efficient as HPA systems.

Also, for this to be truly an aeroponics system the reservoir most be separated from the grow chamber of the plants.

The basic components of a LPA are as follows:

1. High flow water pump

2. Micro sprinklers

Aeroponic systems on the market

Getting back to the definition of Aeroponics: Air at work – This means the plant roots are suspended in midair.

Simply put the plant roots are not covered, not buried, not submerged in any liquid or matter; they are openly suspended in thin air.

So if some or all of the roots are sitting in any liquid, then it's not aeroponics. Any system using a single container to grow and store nutrients, can't be aeroponics.

Aerogarden is a Fake

Aerogarden is Not aeroponics – Ginger Booth, the Indoor Salad Lady, calls it a glorified nutrient film technique hydroponics system. It's sold on the market alluding it's aeroponics. However, it's just a twisted hydroponic system. The plant roots are submerged in a pool of water. Guess what? It's a single container solution.

Tower Garden aeroponic growing system is NOT TRUE

Tower Garden is aeroponics LPA style, it's not True Aeroponics. The system drips the nutrient over plant roots in a hollow tube. I like Tim Blank the inventor. Due to the Tower Garden being easy to operate and includes a very supportive network, it has become popular with many gardeners. However, it's overly sold as THE AEROPONIC system which it's not. The Tower is sold as part of a multi-level-marketing company that became successful providing and healthy choices.

Remember, true aeroponics must atomize the nutrient to a fine mist using high pressure.

How Does Aeroponics Work?

Vegetables out of thin air? Learn more about this environmentally-friendly approach to food production. Towers and other vertical approaches are increasingly popular for aeroponics systems. Since the roots have need to spread out, this is a clever way to save space. A vertical setup also allows misting devices to by placed at the top, allowing gravity to distribute the moisture.

Aeroponic systems nourish plants with nothing more than nutrient-laden mist. The concept builds off that of hydroponic systems, in which the roots are held in a soilless growing medium, such as coco coir, over which nutrient-laden water is periodically pumped. Aeroponics simply dispenses with the growing medium, leaving the roots to dangle in the air, where they are periodically puffed by specially-designed misting devices.

In aeroponics systems, seeds are "planted" in pieces of foam stuffed into tiny pots, which are exposed to light on one end and nutrient mist on the other. The foam also holds the stem and root mass in place as the plants grow.

The Advantages of Aeroponics

Who knew naked roots could survive, much less thrive? It turns out that eliminating the growing medium is very freeing for a plants' roots: the extra oxygen they are exposed to results in faster growth. Aeroponic systems are also extremely water-efficient. These closed-loop systems use 95 percent less irrigation than plants grown in soil. And since the nutrients are held in the water, they get recycled, too.

In addition to these efficiencies, aeroponics' eco-friendly reputation is bolstered by the ability to grow large quantities of food in small spaces. The approach is mainly employed in indoor vertical farms, which are increasingly common in cities − cutting down on the environmental costs of getting food from field to plate. And because aeroponics systems are fully enclosed, there is no nutrient runoff to foul nearby waterways. Rather than treating pest and disease with harsh chemicals, the growing equipment can simply be sterilized as needed.

How Much Does an Aeroponics System Cost?

DIY models can be made for less than $100, but good quality professional systems with automated nutrient monitoring and a backup power supply start in the four-figure range.

Equipment Considerations

All aeroponics systems require an enclosure to hold in the humidity and prevent light from reaching the roots (this is typically a plastic bin with holes drilled for each plant), plus a separate tank to hold the nutrient solution. Beyond these basic components, there are a few other things to consider in devising an aeroponic system to suit your needs.

Some aeroponics systems are designed to be used horizontally, like a traditional planting bed. But towers and other vertical approaches are increasingly popular – since the roots need to spread out, this is a clever way to save space. Vertical systems are also popular because the misting devices may be placed at the top, allowing gravity to distribute the moisture.

Another dichotomy in aeroponic equipment: high-pressure versus low-pressure systems.

Low-pressure systems, which rely on a simple fountain pump to spray water through the misters, are inexpensive and suitable for DIY construction. This approach is sometimes called "soakaponics," as low-pressure misters are capable of producing only a light spray, kind of like a tiny sprinkler, not true mist.

For true mist – meaning moisture floats in the air and more effectively delivers nutrients to the roots – you need higher water pressure than an ordinary pump can provide. Thus, professional aeroponics systems rely on a pressurized water tank capable of holding 60 to 90 psi, along with top-quality misters capable of delivering the finest possible puff of moisture.

Hydroponics suppliers increasingly stock a full-line of aeroponics equipment, from the nutrients, pots, pumps, timers, and tubing you need for a DIY system to fully-automated turnkey aero-farms.

What Can You Grow with Aeroponics?

Anything, in theory. In practice, aeroponics systems are primarily used for the same applications as hydroponics systems, including leafy greens, culinary herbs, marijuana, strawberries, tomatoes, and cucumbers. One exception is root crops, which are impractical in a hydroponic system, but well-suited to aeroponics, as the roots have plenty of room to grow and are easily accessible for harvesting.

Other vegetable crops are possible but have more complex nutrient requirements. Fruiting shrubs and trees are impractical in aeroponics systems due to their size.

A Deep Look At Aeroponics

As our soil quality begins to deteriorate, many people are looking for alternative methods to grow fresh vegetables for their future homes. Because of this reason, hydroponics agriculture will become more popular in the future.

Aeroponics is a simple concept yet it is the most technical among all 6 types of hydroponic systems. Aeroponic system is used by a lot of home growers because it has brought really good results for them.

Aeroponics is actually a subset of the hydroponic system. However, with the hydroponics method, plants use water as the growing medium while aeroponics uses no growing medium at all. This technique was invented during the 1940s and since then, many researchers have added to the theory and application of this method. Aeroponics is considered one of the best methods to grow plants in a soil-free environment and the need for this method has been growing due to a clear need for a more convenient way to grow plants.

In the aeroponic system, plants are not contained in any solid material such as Rockwool or soil. Instead, plant roots are hung in the air in a grow chamber in a closed-loop system. The roots are sprayed with nutrient-rich water or fine, high-pressure mist containing nutrient-rich solutions at certain intervals.

This makes aeroponics a more advanced form than the hydroponic wicking systems, deep water culture, and other types

As all plants need nutrients, the organism will spend a valuable amount of energy growing roots to find these pockets of nutrients in the soil for flower formulation and growth. With hydroponics and aeroponics, nutrients are instead delivered straight to the roots.

Compared to regular hydroponic plants, the plants tend to grow faster and absorb more nutrients because the roots are exposed to more oxygen. Also, there are fewer threats of diseases around root zone disease because there's no place for debris or pathogen to reside.

However, you need to be aware of the fact that aeroponic system chambers are constantly wet with nutrients spray which is convenient for harmful bacteria and fungi to develop. Therefore, it is important to clean and sterilize these misters before using, and occasionally take out and keep these chambers treated with the hydrogen peroxide solution, which can be purchased at any quality hydroponic store.

HOW DOES THE AEROPONIC SYSTEM WORK?

In the aeroponic system, plants are usually inserted into the platform top holes on top of a reservoir and placed within a sealed container.

Due to no root zone media for plants to anchor in, you need to prepare a support collar that will hold stems in place. These collars must be rigid enough to hold plants upright and keep the roots in place but flexible enough to allow room for roots to grow.

The pump and sprinkler system creates vapor (which is a hydro-atomized spray mixture of water, nutrients and growth hormones) out of the nutrient-rich solution and sprays the mist in the reservoir, engulfing the dangling plant roots and absorbed by them. This spray provides the exact amount of moisture which stimulates the plant's growth and allows it to develop turgidly.

The timer supplies the timed spray intervals and duration for the plants. Some people think that growing plants in aeroponic system would be frailer compared to hydroponics. But that's not true. The secret of aeroponic system all lies in the amount of oxygen exposed to the roots without a root zone media limiting it.

Thanks to this, the plant roots will develop rapidly and grow in a moist air-rich environment. If you want to see their development rate, just lift up the growing chamber to see how they are growing.

Types of Aeroponic Systems

Low-pressure Aeroponics (LPA)

This is the most commonly used aeroponic type used by most hydroponic hobbyists due to its ease to set up, availability at any hydroponic shop, and low cost.

Low-pressure creates droplet size much different from the high-pressure aeroponic system.

What you need for this system is just like any hydroponic system – a pump strong enough to move the water onto the sprinkler heads to spray water around the plant root zone.

High-pressure Aeroponics (HPA)

This type of Aeroponics is more advanced and quite costly to set up as it would require specialized equipment. So they are often used in the commercial production rather than home growers.

The HPA must run at very high pressure to atomize water into tiny water droplets of 50 microns or less.

This system creates such a fine droplet size that create more oxygen for the root zone than the LPA, making it the most efficient system among all aeroponic types.

Ultrasonic fogger Aeroponics

Ultrasonic fogger Aeroponics, or commonly called fogponics, is another interesting type of Aeroponic system.

As the name means, growers would use an ultrasonic fogger to atomize water into super small droplets of water. These are very tiny and you will see it in the form of fog.

Though plants roots find it easier to absorb water in tiny size, there's little moisture in the fog created, and when running over time, it can easier create the salt that can clog these foggers than other Aeroponic types.

For more details about this system, please refer to our post about fogponics.

WHICH PLANTS TO GROW?

You can use this system to grow nearly any type of plants and cultivars such as vegetables, nursery stock, houseplants, and bedding. Hundreds of species of plants have been tested and grown successfully by commercial greenhouse owners, researchers and nursery operate using this technique.

TOOLS NEEDED

What you will need:

A reservoir/container to hold the nutrient solution

Nutrient pump

Mist nozzles

Tubing to distribute water from the nutrient pump to the mister heads in the growing chamber

Baskets to suspend plants

Enclosed growing chamber for the root zone

Watertight containers for the growing chamber where the plant's root systems will be

Timer (preferable a cycle timer) to turn on and off the pump

You can purchase these tools at the local gardening/hydroponic supply store or online.

It's quite easy to understand how an aeroponic system works. The purpose of the plant roots hanging in mid-air is to get them exposed to oxygen as much as possible. The high volume of oxygen exposure will stimulate their growth and help them grow faster than they would in the soil, which is a very important benefit of this type of system. This can be seen in the massive root growth of the plants. It has also been proven that this technique increase crop yield X10 compared to using soil. Not only will you be able to collect healthier plants but also grow more crops per year.

No growing media is needed with aeroponics. You can use baskets or closed cell foam plugs (compress around the plant's stem) to suspend the plants. These tools fit in the small holes on top of the growing chamber. In the growing chamber, the mist nozzles will spray the nutrient solution to the plant roots at short intervals. This regular spray has benefits of keeping the roots moist and provide the nutrients they need.

The growing chambers should be airtight and light proof (to prevent the root zone being penetrated by lights and make algae cannot thrive). It must allow air to get in for the growth of plant roots but you also don't want pests to get in or water to spill out. The chamber needs to be able to hold in the humidity as well. The ultimate factor which creates a successful aeroponic system is a balance of plenty of moisture, nutrients and fresh oxygen that you can provide to the roots.

Finally, a major factor which contributes to the success of this system is the water droplet size. A fine mist would create much faster-growing and bushier roots.

These roots also have more surface area to absorb the oxygen and nutrients compared to those sprayed with small streams of water from small mist nozzles. This will also mean faster-growing plant canopy.

The Reservoir

This is where all your water and the nutrient solution will be stored. It's a closed-loop system, meaning whatever the plants don't absorb, gets dropped back into the reservoir to be re-sprayed until the plant roots absorb the water solution. The plants are never submerged in the water though. They're suspended in air using net cups as grow chambers.

Water/Nutrient Pump

Secured to the base of the reservoir is a water pump that's used to pump the water through the piping to the misting nozzles.

Repeat Cycle Timer

The repeat cycle timer is used to control the amount of water dispersed in the reservoir. If you research the recommended cycle times, you'll find growers having success with a misting cycle of one minute on and five minutes off, and others having similar successes on misting for 15-seconds and then off for up to five minutes.

There are no hard and fast rules as to what frequency you set your misting intervals. The important part is that you do set it, because otherwise, the roots will be drenched.

Your best bet is to test every new plant when you start out because the best cycle is the one that lets the plant roots nearly dry out before hitting them with another burst of atomized water.

Misting Nozzles

Different aeroponic systems will have a different number of misting nozzles used inside the chamber. These are an important part of the set up as the smaller the water droplet sprayed through the mister, the better the plant roots can absorb it.

A study by NASA research found that the best range is between 5 and 50 microns for the water droplets, which is the generally accepted standard for a high pressure aeroponics system. The finer the droplets the better the plants can absorb it.

Net Cups / Grow Chambers

Separating the plant roots from the plant tops is done using a lid with precisely cut holes to insert net cups that are used as grow chambers. These cups are inserted through the lid and sealed with (usually) a Styrofoam collar that provides both support for the stems and acts as a water barrier to keep the water contained in the reservoir.

All aeroponic systems have the same components and work the same way, but there are different types…

Fogponics

Fogponics is a more recent advancement in aeroponics that really takes things to another level. Instead of your plant roots being suspended in the air and sprayed with a fine mist, a fogponics system doesn't use a pump; it uses ultrasonic technology.

It's a disc that's submerged in the water and vibrates at extremely high frequencies that turns the water into a gas form getting water micron sizes down to just one micron and often less.

To really comprehend how small that is, one micron is equal to 1 millionth of a meter. In inches, it's 0.00004."

CHAPTER THREE

Growing Hydroponically with Fogponics

Growing with fogponics is the #1 use for ultrasonic mist makers today... There are many different benefits to using fogponics vs. hydroponic methods such as DWC, Aeroponics, ebb and flow. Ultrasonic mist makers create a very small .5 μ (Micron) sized water droplet. This droplet is so incredibly small that it is considered "dry fog". These tiny water droplets easily defy gravity, providing every bit of root area what it craves... 100% coverage, with rich nutrient dense fog.

One of the great benefits of Fogponics over other applications, is that the ultralight fog allows very small hair like shoots to grow out of the main root mass. This provides an exponential surface area for nutrient and oxygen intake providing exponential growth!

Fogponics Systems save money and space!

Hydroponic fogging saves time and money! Using fog over standard hydroponic methods can reduce wasted water and nutrients by nearly 50%! A fogponics system is cleaner, and easier to work with than standard hydroponic systems with messy growing mediums. When growing with a fogponics, roots are suspended in the air allowing them to grow unobstructed and worry free.

A large benefit of fog is that it does not lose its nutrient concentration when cycling. Methods like NFT or nutrient film technique are great, but as nutrient water passes across the root masses, it gradually becomes less and less fortified. As a result, plants at the end of the nutrient water cycle will not benefit as much as plants early on in the cycle. Growing with fog eliminates this problem allowing unused, fully fortified fog to travel throughout 100% of the roots zone.

Whether you have a grow area the size of a warehouse, or the size of a shoebox, fogponic systems can be built to accommodate any need. You can contact us for info on systems with up to 48 discs!

Building a Fogponics setup

A small "fogbox" like Nutramist can be easily built at home using just a few things purchased online.

• 5 gallon bucket and lid

• waterproof micro fan

• six speed variable voltage plug from radioshack

• 1/4" tubing

• 1/4" NPT float valve (small) (not the float you will receive with your purchase- that float is not needed for this project)

• 1/4" barb fittings

• 1-2" flexible hose

• 1-2" barbed bulkhead fitting

• External nutrient reservoir of your choice

1. First, set your mist maker in the bottom of the 5 gallon bucket and fill it with water until it is at its optimal mist producing height. Mark your water height on the bucket and empty everything out.

2. Now you must determine where to drill a hole to install your float valve. This is in order to keep the water level at your predetermined height.

3. After installing the float valve, connect your quarter inch tubing and route it to your main nutrient reservoir. Connect it to this reservoir with your 1/4" barbed fitting.

4. Now you will need to choose a spot to install your fan. Find a spot based on your own needs. Just make sure that you stay away from any spash that the mist maker may put out. when installing the fan, make sure that it is blowing fresh air into the bucket creating a positive pressure zone.

5. Opposite of the fan, you will install your 1 to 2 inch bulkhead fitting along with your output hose. Positive pressure from your fan will force fog up and out of your new fogponics unit.

6. Fog can be routed directly into your root chamber, and then cycled back into the bucket through either a small drain with quarter inch tubing, or a full-size hose similar to your output hose.

Easy ornamental houseplants to grow via fogponics

Two great house plants are being grown via hydroponics methods (soil less and drip irrigation). They should do well with the fogponics approach.

1. Orchids. The "air plants" which includes plants sold as air plants (not needing water but absorb it just via the atmosphere, often seen tacked to a fridge), the bromeliads (water holding cones like the pineapple plant), and orchids:

These plants are the "inventors" of soil less growing: they live in their original environment without soil! The Bromeliads of the Amazon Jungle are rained on daily with both water and dust nutrients blown from the Sahara Desert.

Orchids often like a mist of water daily, but like roots that dry out. One could consider fogging the plant roots, adjusting the electric timer, to fog for 2 hours then go off. That could duplicate how these plants thrive in nature, particularly the orchid strewn Atlantic Rain forest of southern Brazil.

Fir Bark, large perlite, and hydroton expanded clay pellets are used. Bark is most natural. This is not the bark found in landscape bags, but toxin free bark sized just for orchids. Perlite may be a bit "moist" for fogponics orchids. Tough orchid roots seem to plow well though through hydroton. Some use foam peanuts, but I find these too light to work with, and always move in the pots.

2. African violets are being grown in hydroponics units as are other green houseplants. Imagine a nice wall of these in a living room, on the fogponic A shaped unit.

One problem with African violets is that they can grow fast. If they grow too fast, they lose the shape and crown like appearance, putting off too many side plants. Take a sterile knife monthly and cut these off. If you do let them develop, the plant can be sliced in parts for duplication. This likely would occur rapidly in the fogponics unit, so you could fill a wall or pass on small plants to others rapidly. You also could sell the extras. I would imagine African Violets would come to maturity and grow 4 times faster than with soil. So, after you place the plant in the fogponics wall, start cutting off the new starts in 90 days. If you cut and divide, be sure the new section has some roots. Some dust the cut plants with a disinfectant (I believe sulfur dust is used, but check with your garden center for a dust for cuttings. Also, dust the new plant with rooting liquid, dust or gel (butyric acid will triple root growth, thereby tripling plant growth speed. It also helps prevent some shock to the plant).

The roots of African Violets is fibrous and tender. They are a tropical soil plant without deep roots. So, they do not do well in course growing media. While I've seen photos of violets in clay pebbles, the tiny roots don't have enough strength to get through the pebbles. So, options would be:

1. use a gentle plant collar and no net pot. The collar should be loose foam because the plant is tender and easily damaged. Neoprene plant collars, unless cut with scissors, are too firm.

2. use peat moss or sand , but put this and the plant in an adult foot sock or put some loose quilters batting around the plant and media. It's important to accomplish two things in selecting the sock or other holder

a. it should stop all peat or sand from falling into the solution, where it could clog fogger or pump

b. but, it should be loose enough that the roots can pass through to the fog chamber. I've seen some using "embroidery plastic mesh" but this firm plastic won't allow the roots through. Instead of going out, the roots will circle inside the pot making a tangled clump. Loose roots in the chamber will absorb more nutrients. Also, a nylon stocking would simply not let any roots pass through. Check your sock or material and see if a pencil can penetrate it. I tried some fleece material and found it was impenetrable. Checking at Walmart and with helper Sheryl who quilts, a thin "low loft" poly batting was found that worked. There are also many natural fibers, should you want that. Burlap could be great, but sheds some fibers.

Ordinary gardeners perlite the size or rice or aquarium gravel in bags (which is tiny rounded pieces intended to not cut fish….or your violet), not the big gravel chips for yards).

Both plants have specific fertilizer requirements: blooming with tiny amounts of nitrogen. Light requirements of each is a standard day. Full spectrum or more warm lights are good, as both originally came from full light to highly filter light forest zones. African violets were a forest floor plant. Your African Violets will grow too fast and get some distortion on the leaves if there is too much light. Too little light and the violet leaf stems will be too long and break off.

Both plants are really worth a try. See our articles on orchids, the Mericlone Hybrids at Hausserman developed for indoor growing temperatures. Both plants put out a fountain of flowers mostly year around.

DIY Fogponic I

A small fogponics garden for Basil, using an IKEA box, an ultrasonic fogger, and an aquarium air pump.

Materials:

1 IKEA SAMLA box, 11l, black + matching lid

1 ultrasonic fogger (I got mine in a pet shop, they're often used in terrariums)

1 Aquarium air pump, plus tube and air stone

Net buckets

A bag of hydroton, large enough to not fall through the net buckets

Plants (I used basil)

Additionally, you'll need a light source. You could just set this up on your balcony, or use a daylight lamp.

Step 1: Get Some Plant Cuttlings

I wanted to put some basil in my fogponics setup. I had a small basil plant on my balcony, so I cut off some stems, placed them in water, and waited a few weeks until they had grown enough roots to place them in the fogponics garden.

Make sure to cut the stems fairly low to the ground, so you have a lof of stem to work with. You'll need to exchange the water every days, and make sure to not have any leaves in the water. They should get enough light as well, but direct sunlight would probably not be such a good idea.

Step 2: Prepare the Box

Take the lid of your box, and make holes for the net cups. If you have a hole saw with the right diameter, that's probably the easiest ways. I did not, so I drew the circles with a pencil, and drilled lots of small holes on the inside, very close to the line. I then took a box cutter, cut out the inside of the circle, and trimmed the sides so the net cups would sit flush on the plastic, without falling through.

Then, make two holes in the box, right underneath to top rim of the box, one for the air hose, and another one for the cable to the fogger. Depending on your fogger, this hole has to be a bit bigger to fit the power connector of the fogger through it.

Step 3: Plants in Net Cups

When the hydroton comes fresh out of the bag, it's usually still quite dusty and dirty, so it's a good idea to wash it before using it.

Now take a net cup, put in a plant cutling so a bit of the roots come through the bottom, and fill the cup with hydroton to stabilize the plan. Repeat for the other cutlings.

Step 4: Setup Air Stone and Fogger in the Box

Place the fogger in the box, and push its plug and cable through the hole near the rim that we drilled earlier. Don't plugin it in the powersupply just yet.

Take some air tubing, push it from the outside into the box through the other whole, and connect it to the airstone. Cut the tubing to an appropriate length, and connect it to the airpump.

Now, fill your Samla box about half with water, and plug in the fogger. It should start to create a mysterious looking layer of fog on the water. Plug in the airpump, and the fog should disperse a bit more, and you can see air bubbles in the water

Step 5: Put in the Plants

Put your plants (in their net cups) in your fogponics box. If you don't fill all holes for net cups yet, make sure to close them off with foil. Now, it's just waiting until stuff starts to grow.

I put the fogger in my system on a timer, so it runs for 30min, and then stay off for 30min. Just didn't trust it to run all the time, and this seems to work fine. The airpump just runs all the time.

DIY Fogponics II

Now here is where things get interesting... What is fogponics? It is way more simple than you might think. The video above shows you how simple a diy fogponics system really is! There is also a lot of fascinating info on aeroponics in the video as well. You just need to atomize water molecules... with an ultrasonic pulsator. Luckily, they are really cheap and fairly reliable.

Items List:

1. Ultra-sonic Pulsator,

2. 5 Gallon Buket,

3. Rock Wool,

4. Float/Bouy,

5. Timer.

You will also receive a very strong suggestion from me to also purchase a float for the pulsator. The float is important because it will keep the pulsator under the correct level of water. Without it, especially when the plants start to get bigger, the water levels drop quickly and the pulsator dries up and can burn out (believe me I've done it). By adding the float, you reduce your risk of damaging your plants and pulsator, and save time because you won't have to check the water levels nearly as often because as the water goes down the float goes down with the water level; as opposed to having the pulsator in a fixed position or manually monitoring the water level. Believe me, this is a "must-have" for your diy fogponic system:

That's pretty much it. You add water to a reservoir, add an ultrasonic pulsator with a float, cover with a lid, cut holes in the lid and I use rock wool to hold the plants and keep the fog in the container. A 5 gallon bucket is the choice of many, including my first few projects and I still use them for cheap, fast veggies and herbs!

Remember to use a timer:

15 mins on 15 mins off is a good start. I always try to stretch my off intervals as long as the plants can tolerate. If the system is built well, the fog will just linger like a clean, healthy fart; feeding and nurturing your precious

plants. It is very important to keep in mind that the plants' roots need to breath. The holes in the roots that absorb nutrients are between 5-10 microns, which happens to be the same size the water molecules are when they are atomized by the ultrasonic pulsator. This makes fogponics the most efficient means of growing plants and that is why NASA is now developing fogponic systems for their space stations. Enjoy and you're welcome!

The Benefits of Cloning with Fogponics

What's the big deal about cloning plants, and why should you use FogPonics?

These are some questions growers might be asking themselves as they try to understand some of the newest practices and strategies around for nurturing hydroponic plants to maturity.

Cloning is a delicate process, and it definitely takes technique. Unlike with plants that grow from seeds or seedlings, you're having to transplant them without defined root systems, and that makes plants much more vulnerable and open to problems.

Here's some of what we've been hearing from growers about how FogPonics as a foliar feeding method supports clones and helps them to grow into good health.

The Principle of Supplementary Feeding

One of the most fundamental ways that Fogponics helps is that it gives growers another way to nurture plants. Traditionally, all of the feeding was done through the plant roots, through nutrient-rich water placed directly in the sterile media that, in some ways, mimicked the soil conditions for traditional plants.

But FogPonics is a different type of feeding called foliar feeding. In this method, growers actually spray or mist nutrient-rich water directly onto the plants stem and leaves. Small glands called stomata process the nutrients and vitamins the plant system.

So FogPonics is essential when plants aren't able to feed well through their roots, for example, when clones don't have developed root systems yet.

Direct Applications

One of the corollary benefits of FogPonics and other foliar feeding methods is that growers can manage the nutrients better. They can control the nutrient mix in a spray bottle a little bit better than the nutrient rich water coming from a deep water culture reservoir.

Although traditional hydroponics irrigation systems can be fine, FogPonics represents a more direct way to get key nutritional elements to plants.

No Pump

When you talk to growers who have been in business for a while, one of the biggest problems that many of them mention is pump failure. You face all kinds of problems with the engineering of getting a well rated pump where it needs to be. Then, pumps can break down. A manually operated spray bottle never breaks down, and that's another benefit of FogPonics for clones and other types of plants.

Fogponics 101

When it comes to plant cultivation I'm a big fan of any unconventional method or technique that actually works.

Unfortunately, there just aren't too many. Gardening is an established skill; we've innovated quite a bit over the last thousand years!

With that being said, within the realm of hydroponics – a method of soilless cultivation – innovation is ripe.

We've seen the rise of automated hydroponic systems for both small home gardens and even large scale commercial ventures.

In short, fogponics is a sub-technique of aeroponics. This technique uses fog – or very fine water droplets – to grow plants, herbs, and veggies.

Within hydroponics, there are three core growing techniques: Liquid Culture Hydroponics: A technique that uses a nutrient solution culture with no solid medium.

Solid-Media Culture Hydroponics: A technique that uses solid media like heavy pots and bags.

Aeroponics: A technique that uses misters, foggers, and sprayers to supply suspended plant roots with nutrient solution.

Fogponics Design

A normal storage box can be used as the support structure of the unit. Cutting small, 8 cm, holes at the top of the lid is perfect for fitting netted cups filled with some sort of growing medium – like coconut coir, or coco pellets – to hold plant roots.

"Fogponic" Vertical Garden System

The Fogger is Hamburg-based studio, Vakant Design's concept for a highly efficient, indoor-outdoor vertical garden "fogponic" planting system. A nearly self-sustainable alternative to the traditional edible garden and great for those with small urban spaces, the design compliments almost any decor while also helping to purify the air indoors. The nutrient tank gets refilled every four to six weeks.

Efficient Use of Space

Made of ceramic, wood, neoprene, and aluminum, the Fogger utilizes only 3.2 square feet (0,3 m²) of floor space but is capable of producing the same amount of food as a traditional in-ground 32 square foot (3 m²) garden.

Generated by an ultrasonic head, a nutrient-rich fog flows around the plant roots inside the pillar, delivering oxygen to maximize the plants' full growth potential. Through the self-contained system, excess water and nutrients are fed back into circulation so they are lost through seepage or evaporation.

How it Works

1. Insert pre-germinated fruit or vegetable seedlings with neoprene plugs into the pillar.

2. Fill up tank with water using the included organic fertilizer.

The Fogger's cable connects to an external power supply which initiates the automatic cultivation process. The plants receive water and nourishment for up to six weeks via a "fogponic" cultivation technique. The easily installed individual ceramic modules are also simple to maintain as they are dishwasher safe.

Lifting off a single ceramic module to inspect the roots.

Once lid is removed, the fog is allowed to dissipate in order to vie the plants.

The Fogger's lid may be used as a bowl in which to place harvested food.

The Fogger began as an intensive research project about possible cultivation techniques. After several growing trials testing the fogaponic technique for efficiency and suitability, the designers concentrated on a vertical and modular concept. They also endeavored to create a high value piece of furniture that would integrate itself through its attractive design into any residential space, indoors or out.

What Plants Work Best with the fogponics

In theory, you can grow all plants with the Fogponics, yet it's advised that you choose appropriate plants for the best results.

Seedling, clonings

There is an excellent reason why fogponics works absolutely well with seedlings and clonings. Young cloners from cuttings do not have a developed root systems yet, so it's really difficult to moisture and provide nutrients to them. The amounts of water must be kept at an appropriate level. Because too much that can suffocate the cloning. Too little, or the roots may dry out, and sprouts will never spring up.

With the fogs emitted from the fogponics systems, growers provide a constant mix of small weak moisture for cloners at controlled amounts.

Green vegs

The fogponics also works best with green vegs, including lettuces, spinaches, kales, cucumbers, beans and so on.

Herbs

Undoubtedly, most herbs like basil, mint, chives, etc thrive in the fogponic system as it is lightweight, and have a short growing life span.

Pros and Cons of a fogponic system

Pros

Tiny sized droplets. Great coverage

According to the NASA research - Aeroponics for Spaceflight Plant Growth by J.M. Clawson, A. Hoehn, L.S. Stodieck and P. Todd BioServe Space Technologies R.J. Stoner 2000, small droplets tend not to impinge on roots. That answers why aeroponics/fogponics is so effective for clonings because the cuttings, and seedlings are still weak, and these low-pressure fog will not hurt the cuttings and roots.

Also, the hanging roots will get more air than they could in the soil, and in traditional Hydroponic systems like Deep Water Culture, Nutrient Film Technique.

What's more, Fogponics does not deliver much nutrient as the small-sized nature of the fog, which is very important for Seedlings as they do not need many nutrients.

High nutrient concentration

Because of this, the nutrient concentration is not lost when traveling within the reservoir. To see how it is a benefit, let's compare with a Hydroponic system, say the Nutrient Film Technique. In the NFT system, when the nutrient cycle comes and runs across the plant roots, those at the beginning of the tray will get the most nutrient, and it becomes less concentrated at the end of the water cycle.

Easy to clean

The fogponics only has 1 unit to clean rather than lots of heads like in the Aeroponics.

Cons

Heats from the atomizer.

When running continuously for a long time, the atomizer will warm the reservoir. The heat can dry and evaporate the fogs. And your roots can dry out subsequently. If that happens, you need to the cool the temperature down. You can do this by getting a timer and set on and off interval for the foggers. Another way is using some chilling methods like adding some ices into it. If you have more money to spend, you can get a water chiller

Built up salt.

Like the aeroponics system, over time you'll see the salt accumulates in the system. This can clog the foggers. That means it'll need regularly cleaning by using a toothbrush or soak them in vinegar to ensure the fogponics will run effectively.

Susceptible to power outage.

As the advanced form of hydroponics, the fogponics relies on the electricity to run the whole system. Also, the plant's roots are hung freely without submerging into the water like the traditional Hydroponic systems. In this sense, in case the electricity is off, the fog will stop. The plants will not able to get the moisture and nutrients. They will dry out fast and the death can be expected.

High initial cost

This advanced form of Hydroponics will cost you some start-up cost in the beginning.

Growing with fogponic tips:

Because of the heats and the risk of evaporation of the fog, you can use a small fan to regulate the fogs.

Occasionally check the nutrients to ensure the salt is not built up as it will hinder your plants' development.

Most of the time, 1 single-node fogger is enough for starters. But when the sizes of your gardens extend, you'll need to get more.

When growing with seedling/cloners. As plants do not need not a large amount of nutrients at this phase, don't add many nutrients into the reservoir.

CONCLUSION

Even though there are some hurdles like some startup cost, regular maintenance, risks from the system interruptions, etc, Fogponics is still a very effective Hydroponic system.

It is fairly interesting to try for Hydroponic starters and hobbyists. In commercial greenhouses, growers have used aerponics/fogponics for growing plants for years but they will use more expensive misting and foggers for better fog production instead of the common cheap ultrasonic foggers.